SOTHIS PRESS

Sothis Press aims to completely reimagine the public domain book with an emphasis on universal access to captivatingly designed print editions. We believe that technology will enable every single book ever published to be perpetually accessible in beautiful print form. Sothis is beauty in ubiquity.

www.sothis.press
1 Waterhouse Square, London, EC1N 2ST

# Instructions for Observing the Total Solar Eclipse of July 29, 1878

United States Naval Observatory

SOTHIS PRESS

ISBN 978-1-0365-1525-6

*Instructions for Observing the Total Solar Eclipse of July 29, 1878* is in the public domain.

# INSTRUCTIONS

FOR

# OBSERVING

THE

# TOTAL SOLAR ECLIPSE

OF

# JULY 29, 1878.

ISSUED BY THE U. S. NAVAL OBSERVATORY.

WASHINGTON:
GOVERNMENT PRINTING OFFICE.
1878.

# TABLE OF CONTENTS.

These instructions were, by my direction, written by Prof. William Harkness, United States Navy.

Sections I, II, V, and VIII describe such observations as can be made with the apparatus usually possessed by amateurs; while the other sections relate mostly to observations which can only be carried out by persons who are able to command expensive apparatus, and who are skilled in astronomy and physics.

JOHN RODGERS,
*Rear-Admiral, U. S. N., Superintendent.*

# INSTRUCTIONS FOR OBSERVING THE TOTAL SOLAR ECLIPSE OF JULY 29, 1878.

The following suggestions are intended to direct the attention of persons who may witness the total solar eclipse of July 29, and who may desire to co-operate with the United States Naval Observatory, to some of the phenomena which, in the present state of science, it is most desirable should be carefully observed on that occasion.

In writing these suggestions, much assistance has been derived from the instructions issued to the government eclipse expeditions sent out by various nations during the last ten years; and free use has also been made of the reports of these expeditions.

## SECTION I.—OBSERVATIONS WITH THE NAKED EYE.

1. At all places on the central line of the eclipse, the following phenomena will be seen during totality, if the sky is clear, viz: Just as the sun's last ray fades out, some glowing points of light will start into view, hanging upon the edge of the black moon and glistening like rubies; while surrounding the whole will be seen a halo whose mild radiance reminds the beholder of the glories with which the painters of old adorned the heads of saints. The glowing points are the red prominences, and the halo is the corona; both of which are now known to be solar appendages. The corona consists of a comparatively bright inner part lying close to the sun, surrounded by a much fainter mass of luminous matter of vast extent, and generally of most irregular form. Accurate drawings of it will be exceedingly valuable, and fortunately, inaccuracy, such as affects the scientific value of the drawings, can be easily avoided. With this view, persons intending to make sketches should provide themselves with a sheet of paper about 9 inches wide by 12 inches long, having upon it a black disk, 1¼ inches in diameter, to represent the moon, from the center of which straight lines are drawn at angles of thirty degrees, the whole being an exact copy of Plate I. Let a weight be suspended by a string in such a position that the observer can see it hanging over the sun's center; and let the diagram upon which the drawing is to be made be so placed upon any convenient stand that the line marked "top" "bottom" shall be in the plane passing through the observer's eye, the string, and the sun's center, the end marked "top" corresponding to the top of the string. Also, let a lighted lantern be ready to illuminate the paper, in case the light of the corona is insufficient. Finally, two or three minutes before the beginning of totality, let the observer close his eyes and cover them with a handkerchief, so as to make them as sensitive as possible. A friend must be ready to warn him when totality begins, and then, removing the handkerchief, he must rapidly sketch in the outline of the faint outer parts of the corona, taking special care to preserve their correct proportions

relatively to the moon, and above all being scrupulously particular to place the prominently marked branches and rifts in their proper positions. In doing this, great assistance will be afforded by the lines drawn upon the diagram from the center of the black disk. Having completed the outer portions of the corona, the outline of the bright inner part should be sketched; and in doing this it will probably be best to view the eclipse through a spectacle glass of a light green tint. Finally, if any time remains, the whole drawing should be carefully revised; with the aid of an opera-glass, if possible.* It is not worth while to insert the red prominences. The duration of totality being only three minutes in the most favorable localities, it is useless to attempt making more than a single drawing; in fact, such an attempt would show that the work had been so hurriedly done as to deprive it of all value.

Plate II is a drawing of the eclipse of April 16, 1874, given by Mr. E. J. Stone, director of the Royal Observatory at the Cape of Good Hope.† It shows the style in which the drawings should be made; *a, a, a,* being the outline of the faint outer portion, and *b, b, b,* the outline of the bright inner part, of the corona. Probably many parts of the outer corona fade away so gradually that it will be impossible to say precisely where they terminate. This uncertainty should be indicated by a serrated outline, as shown at *a′*. The original sketches must not be altered after the totality is over; but if the observer desires to make a finished drawing from memory, it may be done upon another piece of paper.

2. Portions of the corona may probably be seen for a few seconds, both before and after the totality. Note the exact positions and appearances of these portions, observing particularly whether the parts seen before totality are the same as those seen after totality, and how long they are visible before and after totality.

3. Note the character of the extreme outer boundary of the corona. Does it fade gradually away, or has it a definite edge?

4. The moon's shadow should be bordered by diffraction bands; and some observers claim to have seen them as dark shadows flying over the earth's surface immediately before and immediately after totality. Probably they would be most visible from an eminence commanding an extensive landscape, and it would be a matter of interest to look for them.

5. As the truth of Leverrier's discovery of an apparently unexplained motion of the perihelion of Mercury is now established beyond all doubt, it is important to renew the search for an intra-Mercurial planet or planets. With this view, the parts of the ecliptic near the sun should be carefully examined during the totality of the coming eclipse, and if anything hitherto unknown is discovered, its bearing and distance from the sun should be accurately measured. The bearing can be defined by holding a watch between the eye and the corona, and giving the hour which points toward the object, it being understood that the XII is toward the zenith. The distance should be given in inches and parts of an inch, as measured by a rule held exactly eighteen inches from the eye. To insure the rule being so held a string should be tied to it, in which

* The outer corona is so faint that different persons may see it quite differently. It is for the purpose of diminishing this personality that the use of the opera-glass is recommended.

† Memoirs of the Royal Astronomical Society, Vol. 42, p. 43. Mr. Stone advocated the method of sketching which has just been described.

there is a knot at a point just eighteen inches from the rule. By holding this knot between his teeth, and keeping the string stretched, the observer will be sure that the rule is at the proper distance from his eye, and then he has only to be careful to hold it at right angles to the string. But one eye should be used in making the measures, and no attention need be paid to any object more than five and a half inches from the sun. Mercury will be quite conspicuous at a distance of seven and a half inches, and Mars will be seen at about five and a half inches.

6. Very good observations of the third and fourth contacts, the difference between which is the duration of totality, can be made by the naked eye; the only instrument requisite being a watch having a seconds hand. The importance of such observations and the method of making them are fully explained in Section V.

7. Note what part of the sky is darkest during totality, the form of this dark part, and how the light varies from it to the brighter parts.

## SECTION II.—TO PHOTOGRAPHERS.

Photographs of the corona of very considerable scientific value can probably be made with the apparatus in ordinary use by photographers. With this view, the camera should be fitted with the largest and longest focused *portrait lens* at the operator's disposal, and then, the diaphragm being removed, it should be most carefully focused upon some distant object. The image will be well defined only over a small space in the center of the ground glass, but that is of no consequence. Two or three minutes before totality, some half-sized plates, prepared with the most sensitive chemicals obtainable, should be in readiness for immediate use. The number of these plates should be equal to the number of half minutes in the predicted duration of totality at the place occupied by the operator. For example, if the predicted duration is $2^{m}\ 30^{s}$, then five plates should be prepared. The camera should also be in readiness, directed to the sun, and held there in any convenient manner. As the sun will be high in the heavens, the ordinary camera stand may not be available, but with a little ingenuity some rough contrivance can be made which will answer the purpose.

As many of the plates as possible should be put in holders and laid beside the camera, so that they can be used without delay. Then, the focus of the camera remaining as it was when adjusted upon the distant object, as soon as totality commences, and the operator has assured himself that the image of the corona is in the center of the ground glass, he should insert a plate and expose it. The ground glass should next be replaced, when, owing to the diurnal motion of the earth, it will be seen that the image of the corona has moved away from the center of the field. It should be put back, after which another plate should be inserted and exposed; and this routine should be continued till the totality is over, when the plates can be developed. With a quick-acting portrait lens, the first plate should be exposed three seconds, the next six seconds, the next three seconds, and so on, alternately. With an ordinary portrait lens the exposures should be alternately five and ten seconds; but it must be understood that diaphragms are not to be used in any case. The development must be carried far enough to bring out all the details; but extreme care must be taken not to push it so far as to clog up the delicate outlying portions of the corona. On account of the motion of the earth while the plate is being exposed, the image of the sun—or rather of the moon, which covers the sun—will be a little blurred, and somewhat oval in shape; and to prevent disappointment, it must be borne in mind that the portrait lenses in ordinary use have such short foci that they can give but very small pictures of the moon. The actual size in inches will be very nearly as follows, viz: With a 1-4 lens, 0.04; with a 1-2 lens, 0.07; with a 4-4 lens, 0.09; and with an extra 4-4 lens, 0.14. The diameter of the extreme outlying portions of the corona may be four or five times these figures.

It is hoped that photographers making pictures in this manner will send the original negatives to the Naval Observatory; but, if they prefer to retain them, it is earnestly requested that they will, at least, send positive copies on glass. Paper prints are useless for scientific purposes. Each plate must be accompanied by a statement of the kind of lens used in its production; the focal distance and aperture of that lens; the chemical process employed in the dark room; whether the bath was neutral or acid; how long the plate was exposed; and the means of identifying its top. The latter point is of special importance, and the upper side of every plate must be carefully marked when it is put into the holder, because the character of the corona is such that without this precaution it will be impossible to say certainly which is the top of the negative.

## SECTION III.—THE PHOTOHELIOGRAPH.

It is known that very accurate positions of Mercury and Venus relatively to the sun's center can be obtained from photographs of the latter body taken during transits of these planets; and it may be inferred that equally accurate positions of the moon might be obtained from photographs of the partial phases of a solar eclipse. Unfortunately such is probably not the case. The clearness of the atmosphere, the sensitiveness of the chemicals employed, the length of exposure of the plate, and the hour of the day, all affect the photographic diameter of the sun to such an extent that it is practically impossible to make a series of pictures in which it shall be even approximately constant. In the case of transits of Mercury or Venus this is of no consequence, because it may be assumed that the change of diameter takes place uniformly all around the circumference; and having a complete outline of the sun, it is always possible to find its center with accuracy. In a similar way it is possible to find the center of the planet, whose diameter is also variable, and thus the positions of the two bodies relatively to each other become known. When the diameter of a celestial body is unknown, at least a semi-circumference of its photographic image is necessary from which to find the exact position of its center. In the case of a solar eclipse the diameters of the sun and moon are nearly equal, and therefore any picture which shows a large part of the circumference of one of these bodies necessarily shows but a small part of the circumference of the other. For that reason it seems impossible to find accurately the center of more than one of the two bodies; and consequently it seems impossible to find accurately the positions of the two bodies relatively to each other. On the other hand, contact observations afford a very accurate means of finding the errors of the lunar tables; and as such observations can be made with ordinary instruments, at a comparatively trifling cost, it does not seem wise to incur the trouble and expense of erecting more than a single experimental horizontal photoheliograph specially to observe an eclipse. At places where such instruments are already in position it may be well enough to use them, but it is extremely doubtful if the pictures will be of much value.

## SECTION IV.—THE EQUATORIAL CAMERA.

Good photographs of the corona will be exceedingly valuable, but they require a special form of apparatus for their production. That best adapted to the purpose is a camera, fitted for using either large or small plates, and furnished with a quick-acting objective, the whole being equatorially mounted and driven by clock-work, so that long exposures can be readily made. To resist the strain consequent upon the manipulation of the plates, the mounting and driving clock must be of the most substantial kind; and to depict the fainter parts of the corona the objective must be as rapid as possible. In the latter matter our choice will be practically restricted to the lenses named in the following table, which embraces all the most rapid varieties in the market. The names of these lenses, and the corresponding intensity ratios, are varied somewhat by the different makers. Those given in the table have been taken from Dallmeyer's catalogue.

| Reference No. | Description of Photographic Objective. | Intensity Ratio. | Focal Distance of Largest Lens Made. | Diameter of Image of Sun. | Exposure required for the Corona. |
|---|---|---|---|---|---|
| | | | *Inches.* | *Inches.* | *s.* *s.* |
| 1 | Extra Quick Acting Portrait . . . . . | $\frac{1}{2}$ | 5½ | 0.051 | 0.3 to 1.6 |
| 2 | Quick Acting Portrait . . . . . . . . | $\frac{1}{3}$ | 13½ | .126 | 0.7 to 3.6 |
| 3 | Ordinary Portrait . . . . . . . . . . | $\frac{1}{4}$ | 24 | .224 | 1.3 to 6.4 |
| 4 | Portrait and Group (D) . . . . . . . | $\frac{1}{6}$ | 33 | .308 | 2.9 to 14.4 |
| 5 | Rapid Rectilinear . . . . . . . . . . | $\frac{1}{8}$ | 33½ | 0.313 | 5.1 to 25.6 |

If $F$ is the equivalent focal distance of a photographic objective; $d$ its working aperture; $C$ the exposure constant, whose value depends upon the intensity of the light and the sensitiveness of the chemicals employed; and $t$ the time of exposure required to produce a good negative; then, the intensity ratio is $\frac{d}{F}$, and

$$t = C\left(\frac{F}{d}\right)^2$$

The data from which to determine an approximate value of $C$ for the corona are very limited. In August, 1869, Mr. J. A. Whipple obtained a negative of the corona at Shelbyville, Ky., by placing the sensitive plate at the principal focus of an ordinary astronomical telescope. In this instrument the value of $F \div d$ was 16.4, and as the exposure was 40 seconds it results from the formula above that $C = 0.149$. In

December, 1870, Mr. A. Brothers, at Syracuse, Sicily, obtained a very thin negative of the corona, with a Dallmeyer rapid rectilinear lens, in 8 seconds. For this objective $F \div d = 8$, and consequently, $C = 0.125$. During the same eclipse, Mr. O. H. Willard, at Jerez, Spain, obtained a negative of the corona in 90 seconds. He used an astronomical telescope, precisely as Mr. Whipple did in August, 1869. If we assume $F \div d$ to be 15, as is usual in such instruments, then $C = 0.400$. The negative did not seem overexposed. In December, 1871, Mr. J. Boesinger, at Ootacamund, India, obtained a negative of the corona in 3 seconds with a Dallmeyer No. 6 D portrait and group lens. In this case $F \div d = 6$, and, therefore, $C = 0.084$. In August, 1869, Dr. Edward Curtis, at Des Moines, Iowa, used values of $C$ which were respectively 0.016 and 0.011; but on that occasion the actinic force of the sun was diminished from ten to twenty times by a dense haze, and therefore these values were not equivalent to more than 0.0016 and 0.0011 in an ordinarily clear atmosphere. With the first exposure the prominences and the lowest layer of the chromosphere were well shown. With the second exposure the prominences were still depicted, but the chromosphere only appeared around part of the sun's circumference. The corona, properly so called, was entirely absent from both pictures.

From all these facts it is probably safe to conclude that, with a clear sky and a moderately high sun, exposures in which the value of $C$ is about 0.002 will give only the prominences and the outline of the moon. When $C$ becomes 0.08 the corona will begin to appear, and will increase in extent as the exposure increases, at least up to the point where $C$ becomes 0.40. Beyond that I am not aware that we have any experience. Accordingly, the shortest exposure specified in the table above corresponds to $C = 0.08$, and the longest to $C = 0.40$.

In choosing a lens from those mentioned in the table, Nos. 1 and 2 may be at once dismissed from consideration, as giving too small images of the sun. Nos. 4 and 5 are practically of the same focal length, and as both of them are entirely free from distortion and flare, No. 4 should be preferred on account of its larger aperture. Thus our choice is narrowed down to Nos. 3 and 4, No. 3 having the greater value of $F \div d$, but neither so large nor so flat a field as No. 4. If the lens were required solely for photographing the corona, it might be difficult to choose between these two, as either of them will cover a plate far larger than can ever be required for that purpose; but it will presently appear that there are questions in eclipse photography for the solution of which plates of the largest size must be used, and on that account it is clearly best to choose No. 4.

Having adopted a lens of 33 inches focus, we have next to consider what size of plates should be used with it. The diameter of the sun's image will be 0.31 of an inch, and, taking Mr. Brothers's picture of the eclipse of December, 1870, as a guide, the diameter of the corona may be five or six times as great, or, in other words, about 1.7 inches, to receive which a plate measuring 4¼ by 5½ inches will be ample. But an attempt should be made to use the lens for another purpose, even more important than photographing the corona. The magnitude of its intensity ratio enables it to depict faint objects rapidly, and the extent of its angle of view is such as to embrace a field of more than forty degrees. It, therefore, seems peculiarly fitted to assist in the search for intra-Mercurial bodies, provided the illumination of the sky does not prove too great. Mercury's maximum elongation from the sun scarcely reaches twenty-nine degrees, and, if we postulate Bode's so-called law, the greatest elongation of the hypothetical intra-Mercurial planet or planets should not exceed fifteen degrees. Double this distance is the space along the ecliptic which must be searched; and to photograph it a plate 17.7 inches long will be required. Our lens will cover a plate measuring 20 by 22 inches, but it is very desirable to keep the apparatus as light as possible, and on that account it seems best to have a camera twenty inches square, and to use plates measuring 17 by 20 inches, which will suffice to cover a space of thirty-three and a half degrees along the ecliptic.

Assuming the adoption of an equatorial camera twenty inches square, provided with a lens whose intensity ratio is one-sixth and whose focal distance is about thirty-three inches, it yet remains to consider how this apparatus should be managed during a totality lasting only three minutes. As the illumination of different parts of the corona varies greatly, there can evidently be no certainty of getting all the details of the phenomenon unless a series of plates are taken, in which the exposures vary from the shortest possible up to the point where it is certain that an increase of time does not improve the picture. On this account it will be desirable to take as many as six plates, the exposures being, respectively,

$$3^s, 5^s, 10^s, 20^s, 40^s, \text{ and } 60^s.*$$

The first four of these plates will receive such short exposures that it is unlikely they will show anything but the corona, and therefore their size should be 4¼ by 5½ inches. With the last two plates the case is different. Their size should be 17 by 20 inches, because their longer exposures will probably suffice to bring out upon them any bright points which may exist within their field. A lens such as is here under consideration should depict an eighth-magnitude star in about one minute, but of course the intensity of the sky-illumination during totality will determine the limit of brightness at which faint luminous points will cease to impress themselves upon the negatives, and what this limit may be it is impossible to predict. The necessity for at least two large plates is evident when it is remembered that the image of a small bright point could not be distinguished from an accidental blemish in the film, and it would only be by finding it upon both plates that its true character could be unmistakably recognized. It is exceedingly desirable to determine accurately the maximum exposure that the corona will bear with advantage, and it is hoped that on at least one of the large plates it will prove to be over-exposed.

For the management of the equatorial camera two persons will be required, designated respectively as No. 1 and No. 2, and the following routine will be found convenient. The operations in the dark room must be so managed that the six sensitive plates shall be placed beside the camera, in readiness for instant use, a minute or two

* It is questionable whether or not all these exposures can be made in three minutes, but a few preliminary trials will settle that point. Experiments should also be made upon the stars in the evening twilight to ascertain if an exposure so long as sixty seconds is really required. In any event, it will probably be best to reverse the order given above, and to work off first those plates requiring most time, thus reducing to a minimum the damage which would accrue from the sun bursting out while an exposure was in progress.

before the predicted time of second contact. Six sand-glasses must also be provided, each adjusted to run for the number of seconds that the plate to which it belongs is to be exposed. Then, the cap being on the lens, the first plate must be placed in the camera, and the slide must be withdrawn from the holder; after which operator No. 1 will take position with his hand upon the cap covering the lens, and No. 2 will seize the proper sand-glass and hold it in such a way that the sand shall all be in the bottom of the glass. At the instant totality begins, No. 1 will say, "Ready?" No. 2 will answer, "Ready." No. 1 will begin the exposure of the plate by carefully removing the cap from the lens, exclaiming as he does so, "Turn;" when No. 2 will instantly turn the glass. Both operators will stand motionless while the sand is running, No. 2 watching carefully for the escape of the last grain, when he will exclaim, "Out," and No. 1 will instantly terminate the exposure by capping the lens. Without a moment's delay No. 1 will push the cover into the plate-holder, remove the exposed plate, place the next holder, containing a fresh plate, in the camera, and draw the slide; being assisted in all this by No. 2, if necessary. Then No. 1 will again take position with his hand upon the cap of the lens; No. 2 will seize the second sand-glass; and this plate will be exposed in the same manner as the preceding. This routine will be continued till all the plates have been exposed, and after totality is over they will be developed in the usual way.

With regard to the plate-holders, it may be remarked that the last two plates must necessarily be in holders as large as the camera, but the first four plates can be more conveniently handled if they are in half-sized holders. To render such an arrangement possible a reducing frame must be provided, which, upon being applied to the camera, will reduce its opening to that proper for receiving the small holders.

As to the photographic process to be employed, if only small pictures are desired, dry plates, prepared with washed emulsion, are recommended; but if any large plates are to be taken the wet collodion process will be preferable. In either case it is specially important to use well bromized collodion, because the bulk of the coronal light has probably a wave length of 5315.9 (the 1474 line), and to such light silver bromide is far more sensitive than either silver chloride or silver iodide. For wet plates there is reason to think that fineness of deposit and richness in details can be best secured by a free use of sugar in the iron developer.

To facilitate the study of the red prominences, it has been proposed to photograph them on a scale of ten seconds of arc to a millimeter. The optical apparatus for the production of such pictures must have an equivalent focal distance of 2062.7 centimeters, or 812.1 inches, and if we take $C$ equal to 0.002, which is probably very near the truth,* the values of $t$ for lenses of various apertures can be found from the formula already given, and will be as follows:

| Aperture of Objective. | $\frac{F}{d}$ | $\left(\frac{F}{d}\right)^2$ | Exposure Required. | Motion of Moon. |
|---|---|---|---|---|
| *Inches.* | | | *s.* | " |
| 6 | 135.3 | 18306 | 36.6 | 20.1 |
| 8 | 101.5 | 10302 | 20.6 | 11.3 |
| 10 | 81.2 | 6593 | 13.2 | 7.2 |
| 12 | 67.7 | 4583 | 9.2 | 5.0 |
| 15 | 54.1 | 2927 | 5.8 | 3.2 |
| 20 | 40.6 | 1648 | 3.3 | 1.8 |
| 26 | 31.2 | 973 | 1.9 | 1.0 |

The last column of the table gives the moon's motion during an exposure of the length specified in the preceding column. A prominence one minute high is a large one, and yet it could scarcely be photographed with a six-inch objective, because twenty seconds of its height would be covered by the advancing moon before the exposure was over. In short, it does not seem possible to photograph prominences during eclipses on the scale here contemplated with an aperture much less than ten inches, and twelve would be far better. Under these circumstances the practicability of the scheme is questionable.

## SECTION V.—TELESCOPIC OBSERVATIONS.

*Telescopes.*—A proper telescope for observing the eclipse is very desirable, but any telescope at all is better than none; and owners of small spy-glasses, or opera-glasses, should bear in mind that their instruments are capable of doing good service. Opera-glasses can be held in the hand with sufficient steadiness, but spy-glasses, however small, will require a stand to produce the best effect. This need not be an elaborate affair, but may be made of wood in the simplest manner, provided only that it is capable of holding the telescope steadily pointed in any required direction. Some artificial support is absolutely necessary, and, in default of anything better, the telescope might even be lashed to a round post.

*Shade Glasses.*—In anticipation of the late transit of Mercury, several letters were received at the Naval Observatory inquiring about shade glasses, and on that account it seems desirable to notice the subject here. The best form of shade is undoubtedly a long wedge of colored glass, achromatized by a similar wedge of white glass firmly cemented to it. The most usual, and by far the most objectionable, color for a shade is red. Either neutral tint or green is excellent, but it is difficult to obtain the latter sufficiently dark. Deep blue has been advocated by some observers, and deep yellow was much used in Germany, but is now apparently losing favor. Shades consisting of a single piece of thick glass are very apt to split, and thus endanger the observer's

* The exposures given by Mr. De la Rue to his pictures of the eclipse of July 18, 1860, indicate a value of $C$ somewhere between 0.0007 and 0.0012; while Dr. Curtis's exposures on the eclipse of August 7, 1869, give $C = 0.0016$.

eye, if the rays of a bright sun are concentrated upon them by an objective of large aperture. To avoid this, such objectives must have their aperture reduced by a cap, two inches being the proper size for a telescope of thirty inches focus, and three inches for one of five or six feet focus. By making the shade of three thicknesses of glass, the piece next the eye being thickest and darkest in color, the other pieces being successively thinner and lighter in color, and all being fitted loosely into their cell so that they can expand freely, it will be possible to use with safety an aperture of five inches upon a telescope of six feet focus. When no proper shade is at hand, an excellent substitute can be made by smoking a slip of thin flat glass, about an inch wide and four inches long, in such a way that its tint may increase uniformly, from very light at one end to a depth at the other end sufficient for viewing the brightest sun without inconvenience. The smoked surface should be protected by a covering glass, of the same size as the original slip, and separated from it by bits of thin card inserted at the ends, the whole being bound together by a narrow strip of paper pasted around the edges. For observing the second contact of the eclipse, a neutral tint, or green, spectacle glass will be found convenient.

Of course it will be understood that the necessity for caps to diminish the aperture is confined to common telescopes. Instruments intended for observing the sun are always provided with arrangements for getting rid of the excessive light and heat without diminishing the aperture, and often without using shades

*Heating of Telescope.*—When observing a bright sun the eye end of the telescope soon becomes quite hot, and thus air currents are set up in the tube which interfere seriously with the definition. To remedy this the tube must occasionally be cooled, either by shading the objective or by turning the telescope away from the sun. As it is important to have the best possible definition while observing the contacts and totality of the eclipse, special care must be taken that the temperature of the tube at these times is the same as that of the surrounding air.

*The Contacts.*—The first touch of the moon upon the sun's limb is technically called the first contact, the complete extinction of the last ray of sunlight is simultaneous with the second contact, the reappearance of the minutest possible portion of the sun marks the third contact, and the disappearance of all trace of the moon from the sun's disk constitutes the fourth or last contact. The first and fourth contacts are sometimes called the external, and the second and third the internal, contacts. The observation of these contacts consists in noting the exact instants of their occurrence, and for that purpose a time-piece with a seconds hand and a telescope are usually employed, but if the latter instrument is not obtainable, the internal contacts can be well observed with the naked eye.

There is generally some uncertainty about the first contact, because the only evidence of its occurrence is a slight indentation in the sun's limb, which, unless the observer knows precisely where to look for it, naturally attains some size before it is seen, and thus the contact is noted too late. If the telescope has a position-micrometer, its wire may be set at the computed point of first contact, and then there will be no difficulty; but in the absence of such an arrangement, the only way is to use a power of not more than sixty diameters, and to watch carefully a portion of the sun's circumference large enough to include with certainty the point in question. The shade-glass employed should be sufficiently dark to destroy all glare, and to permit of the sun's image being viewed without effort.

The internal contacts are phenomena of such a definite character that the instants of their occurrence can be noted within a small fraction of a second. For a minute or two before the predicted time of second contact the sun's thin, and now fast waning, crescent should be carefully watched through a neutral tint or green shade glass as light as the eye can conveniently bear. Presently the crescent will become a mere thread of light, which will rapidly shorten, and suddenly disappear. This is the second contact, and there can be no mistake about the exact instant of its occurrence because the ruddy glow of the chromosphere, which remains at the point last covered by the moon, is totally different from the dazzling white light of the photosphere which has just disappeared. Through a red shade glass the contrast between these two kinds of light might be partly lost, but through a neutral tint or green glass it will be sufficiently apparent. Finally, there is yet another indication of the second contact, and that is the seemingly miraculous appearance of the complete outline of the moon, round and black, reposing upon the wondrous radiance of the corona.

The approach of the third contact will be heralded by the rapid brightening of the chromosphere at the point of the moon's limb where the sun is about to reappear, and two or three seconds later a sudden burst of light will announce the contact itself and with it the termination of totality. This contact may be observed without any shade glass.

The observation of the fourth contact is a very simple matter. The segment cut out of the sun by the retreating limb of the moon is carefully watched as it becomes less and less, and the instant of its final disappearance is noted as the fourth contact. The shade glass used must be so dark as to destroy all disagreeable glare.

To observe the contacts, a station must be chosen where there is an uninterrupted view of the sun, and the party must consist of only two persons, one of whom must have a time-piece and writing materials and the other must have a telescope. With regard to the latter instrument, it may be remarked that if it is a common one—a spy-glass, for instance—there will probably be no choice as to its magnifying power; but if it belongs to the better class, and has a battery of eye-pieces, the power chosen should be equal to fifteen or twenty times the diameter of the objective in inches. To secure the best results, the party must be entirely alone; and above all, it must not be within hearing of other observing parties. A minute or so before the expected time of contact the person with the time-piece will fix his eyes upon its seconds hand and begin silently to count the seconds, while the observer at the telescope will carefully watch the sun. Upon the occurrence of contact the observer will cry "Mark," and the person at the time-piece will immediately write down, first, the second; then, the minute; and, finally, the hour. Before beginning to observe, care must be taken to see that when the seconds hand points to 60 the minute hand is exactly over a minute mark; and in observing, if the seconds are near sixty, special care will be required to avoid recording a wrong minute. For example, when the time is $4^m\ 58^s$ the minute hand will be almost exactly over $5^m$; and special care must be taken not to record $5^m$ $58^s$ instead of $4^m$ $58^s$. Practice in calling and noting time previous to the eclipse will greatly diminish the risk of such mistakes. A little before totality a lantern should be lighted, because without it there may be difficulty in seeing the hands of the time-piece at the approach of third contact.

Observations of the internal contacts can be utilized without a knowledge of the error of the time-piece; but with the external contacts the case is different. They will be of no value unless the error of the time-piece is accurately known, either on local or on Washington time. To meet this difficulty, it is hoped that arrangements may be made by which the Western Union Telegraph Company will furnish, free of cost, a signal from the Naval Observatory, indicating the instant of Washington mean noon, to all proper persons who may apply for it at any of their offices on July 27, 29, and 30. The intending observer can best utilize these signals by making sure that the minute and seconds hands of his watch agree with each other, as explained above, and then carrying the watch to the telegraph office on each of the days mentioned and noting the hour, minute, and second indicated by it when the signal is received. No matter how the watch may run, it must not be meddled with in any way, either by moving its hands or touching its regulator, between July 26 and 31. Without this precaution, neither its rate, nor the degree of dependence to be placed upon it, can be determined.

The final record of a set of observations of contacts should contain the following information:

1. The name of the station; including town, county, and State.
2. The date.
3. A description of the location of the station with reference to the nearest prominent landmarks; as, for example, its bearing and distance from the nearest court-house, railroad station, or church; or its position with respect to the lines of the United States land surveys; or the bearing and distance of a prominent mountain peak; or the bearings of three or more such mountain peaks; or, in the case of a city, the street and number, together with the names of the nearest cross streets on each side.
4. A description of the time-piece, and a statement of the aperture, focal length, and magnifying power of the telescope employed.
5. The times indicated by the watch at the reception of the Washington noon signal on July 27, 29, and 30.
6. The time of each of the contacts; just as read from the face of the watch, without the application of any correction.
7. An estimate of the uncertainty to which these times are liable.
8. Any remarks which may be thought necessary.
9. The signatures of the two observers, thus: A. B., observer with telescope; C. D., time observer.

It is particularly requested that the original pencil memoranda of the times of contact be sent to the Naval Observatory, inclosed in the above-described report.

If several parties observe the eclipse in the same neighborhood, it is quite likely that their recorded times may differ a little. Such differences occur even in the work of the most practiced astronomers; and they must be allowed to remain. It is never permissible to alter the record. A mere suspicion that such a thing has been done will insure the entire rejection of the observation.

*Limits of the Shadow Path.*—There is yet so much uncertainty in the best solar and lunar tables that in the eclipse of July next the British Ephemeris places the path of the moon's shadow, while crossing Colorado, about three and a half miles southwest of the position assigned by the American Ephemeris. It is therefore important to determine accurately the true position of the path, and this may be readily accomplished by observing the duration of totality at points situated from one to ten miles within the shadow. As the duration of totality is the interval between the second and third contacts, it is determined by observing these contacts in the manner already described; except that for this purpose a knowledge of the error of the watch is unnecessary, and the observations may be made with the naked eye if a telescope is not available. At places very near the edge of the shadow the totality may be so brief that the time observer would risk losing the third contact if he stopped to record the second. At such places it will be necessary to have a third person to make the record, and then both contacts can be readily observed, even if they follow each other at an interval as short as a single second.

*Sketches of the Corona.*—In making a general sketch of the corona the magnifying power employed should not exceed from twenty to forty diameters, and some mechanical means should be adopted to insure accuracy of outline. Owing to the faintness of the corona, it is doubtful if the camera lucida can be successfully used; and the most promising plan seems to be a system of squares, ruled on glass, in the focus of the eye piece of the telescope, and a corresponding system drawn upon the paper on which the sketch is to be made. It is known that spider lines in a micrometer sometimes cease to be visible under the feeble illumination of the corona, and to avoid the possibility of this happening with the squares, the lines defining them should be rather coarsely cut. More than once attempts have been made to use a telescope after the manner of a camera obscura, by drawing out the eye piece a little, and thus causing it to throw an enlarged image of the corona upon a piece of finely ground glass, or semi-transparent paper, placed a few inches beyond; the whole being so arranged that the image could be seen through, and traced upon, the glass or paper. Owing to the faintness of the image, such attempts have not hitherto been very successful, but perhaps they might do better if for the plate of ground glass a thin sheet of gelatine, or mica, were substituted, and the drawing were traced upon it with a needle point. The proper mode of executing such sketches has been already described in Section I.

*Minute Examination of the Corona.*—With as high a magnifying power as the telescope will readily bear, examine the structure of the corona; noting particularly whether it is homogeneous, nucleated like a star cluster, or filamentous. If any nuclei are seen, describe their shape, size, color, and mode of distribution. If filaments are discerned, state their length, diameter, color, and mode of distribution; whether they are straight, curved, or contorted; in what direction they lie relatively to the sun; and, if curved, specify how the curvature faces. Be sure to observe how the structure varies in passing from the sun to the outermost visible limit of the corona.

Since the spectroscope furnishes an efficient means of studying the red prominences at any time, it will be very undesirable to waste a single one of the precious moments of totality in examining them.

*Intra-Mercurial Bodies.*—To facilitate the work of such astronomers as may desire to search for intra-Mercurial planets with considerable telescopic power, a chart will be found at the end of these instructions showing every star so large as the seventh magnitude in that portion of the heavens which will be occupied by the sun on the 29th of July next. The black circle in R. A. $8^h\ 36^m$, Dec. $+ 18° \ 39'$ indicates the position of the sun. Mercury, Regulus, and Mars will be pretty close together, and probably quite conspicuous during totality, but they are so far to the eastward that only the last-named comes within the limits of the chart. Venus may also be seen, but she will be low in the western sky. While looking for planets, the possibility of discovering a small comet, or a meteor stream, should be borne in mind.

*Partial Phases.*—The corona forms a luminous background upon which the moon's limb is sometimes seen projecting beyond the sun; and a little before totality it is even possible that the complete outline of the moon may become visible. Look for these phenomena, and note the time of their occurrence. It is difficult to assign any reason for the existence of rays, or brushes, of light at the cusps of the sun, but it is said they have been seen. If any such appearances present themselves, they should be carefully scrutinized to ascertain if they change either their position or intensity; and the interior of the telescope should be examined to make sure that they do not originate in reflections, either from the tube, or from the lenses.

## SECTION VI.—SPECTROSCOPIC OBSERVATIONS.

This section has been largely compiled from the writings of Young and Lockyer, some of the paragraphs being almost in the exact language of these authors.*

Among the numerous arrangements of spectroscopic apparatus devised by scientific men, experience has shown that those mentioned below are best adapted to the observation of eclipses, viz:

*Analyzing Instrument, A.*—A spectroscope having a long slit and a long collimator, mounted at right angles to the optical axis of a reflecting telescope of large aperture and short focus; the slit of the spectroscope lying in the focus of the mirror as usual and the telescope being provided with a large finder. In this combination the image of the corona is very bright, and small enough in proportion to the length of the slit to allow of the simultaneous visibility of the spectrum on both sides of the dark moon; while the long collimator permits the use of a wide slit, and thus insures the admission of the maximum of light.

*Integrating Instrument, B.*—A spectroscope, having a long collimator, mounted so that the light falling upon its slit is received from the eye-piece of a telescope of large aperture and short focus, previously adjusted for distinct vision on a distant

* See the following: Report on Observations of the Total Solar Eclipse of December 22, 1870. By Prof. C. A. Young U. S. Coast Survey Report for 1870, p. 141. Instructions for Observers at the English Government Eclipse Expedition, 1871. Nature, 1871, vol. 4, p. 517. English Government Eclipse Expedition, 1875. Instructions to Observers. Nature, 1875, vol. 11, p. 351. Report on the Total Solar Eclipse of April 6, 1875. By J. N. Lockyer and Arthur Schuster. Phil. Trans. The Coming Total Solar Eclipse. By J. N. Lockyer. Nature, 1878, vol. 17, pp. 481 and 501.

object. The area of the heavens from which this telescope collects its light should be three or four degrees in diameter, which will be the case if it magnifies about fifteen times. To insure accurate pointing it must be provided with a finder.

*Analyzing Instrument, C.*—An automatic spectroscope of large dispersive power; attached to an equatorial telescope in the usual manner.

*Analyzing Instrument, D.*—A spectroscope of moderate or small dispersive power; attached to a telescope as above described.

*Integrating Instrument, E.*—A spectroscope, either mounted upon a stand or held in the hand, receiving the light of the corona directly upon its slit without the intervention of any optical apparatus whatever.

*Integrating Instrument, F.*—A meteor spectroscope, having neither slit nor collimator, but receiving the light directly upon its prism.

The telescopes to which the above-described spectroscopes are attached should all be equatorially mounted; and they must either be driven by clock-work, or an assistant must be employed to point them. When a spectroscope slit is covered from view the only available mode of bringing an object upon it is by means of the finder; but if it is exposed, as it should always be, the image of the object upon the slit-plate will be the best guide. For observing the corona the width of the slit may be such as to show but two of the three *b* lines when the spectroscope is directed to a faint cloud. The instruments best adapted for making the observations suggested below have in every case been indicated by the use of the reference letters employed in the descriptions of apparatus just given.

*Brushes of light at cusps.*—Some observers report having seen brushes of red light at the solar cusps during the partial phases of eclipses. It is difficult to understand how such phenomena can occur, but if they present themselves their spectra should be carefully examined. Instrument, *C* or *D*.

*Young's Reversing Layer.*—Close to the sun's limb look for the layer in which the Fraunhofer lines originate. Just before totality it should give a nearly continuous spectrum, and just at the instant of totality it should show reversed all the dark lines of the spectrum, except those of terrestrial origin. The dispersive power of the spectroscope employed cannot be too high. The image of the sun should not be less than one inch in diameter, and since the thickness of the layer in question does not exceed $2''$, it is evident that extreme care must be exercised in adjusting the focus, in making the slit very narrow, and in placing it precisely in the plane of, and rigorously tangent to, the solar image. Instrument, *C* or *D*.

An attempt should be made to measure the thickness of the reversing layer by noting how long the lines continue bright. But for this purpose a chronograph capable of indicating the one hundredth part of a second will be necessary.

*The Prominence Spectrum.*—Just before and just after totality the atmospheric glare is greatly reduced, and advantage should be taken of this circumstance to look carefully for new lines. The search should be directed particularly to that part of the spectrum lying above *F*, as it deserves, and will probably repay, careful study, and is very difficult to deal with in the absence of an eclipse. Instrument, *C*.

During totality, attend to the following points:

In passing outward from the sun, are the *b* lines replaced by a diffuse band? Does the *F* line extend higher than the other hydrogen lines? Instruments, *A*, *C*, *D*.

It is known that the prominences consist mainly of hydrogen; but do they ever contain large quantities of calcium? In other words, are the *H* lines sometimes conspicuous in them? Instruments, *A*, *C*, *D*.

Probably the *h* line is produced only by very hot hydrogen. As giving a clew to the temperature of the prominences, look for it in their spectrum, and note how high above the photosphere it exists. Instruments, *A*, *C*, *D*.

Look for new lines in the ultra-violet part of the spectrum. Instrument, *C* or *D*, provided with a fluorescent eye piece.

*The Corona.*—Direct attention specially to the following points:

Does the line 1474 stop short of the chromosphere, or does it enter that formation? If the latter, does it broaden out as it nears the sun? Instruments, *A*, *C*, *D*.

Are there any parts of the corona which give a spectrum containing bright lines between $D_3$ and 1474? In trying to answer this question use rather a wide slit, and, if any lines are found, locate the part of the corona giving them. Instrument, *C* or *D*.

Besides the well-known bright line 1474, the corona gives a faint spectrum, which most observers have thought to be continuous. Is it really continuous, or is it crossed by the Fraunhofer lines? Look particularly for *D*, *E*, *b*, and *F*, and if any of them are seen note in what part of the corona. Instrument, *C* or *D*.

Examine the dark rifts in the corona very carefully to make sure whether or not they give any spectrum. Instrument, *C* or *D*.

Does the corona give any ultra-violet lines? Instrument, *C* or *D*, provided with a fluorescent eye piece.

Examine the brightest parts of the outer corona in order to determine how far from the chromosphere any spectrum is visible, and note also what that spectrum is. Instrument, *C* or *D*.

Examine what part of the coronal spectrum is polarized. This may be done by noting the effect produced upon the spectrum when a Nicol's prism placed before the slit is rotated; but the results thus obtained will be complicated by the polarization due to the passage of the light through the dispersing prisms, and to eliminate this the slit must be placed at different angles to the vertical, and upon various parts of the corona. The following seems a more satisfactory mode of proceeding: If it is borne in mind that the polarization of the corona is partly radial to the sun and partly in the plane passing through the zenith, the sun, and the observer's eye, it will be evident that the parts of the corona cut by a vertical circle passing through the sun's center are polarized only in the plane of that circle. Hence, if the slit of the spectroscope is placed in that plane, and the image of the sun is allowed to fall upon it in such a way that the sun's center coincides with the axis of the slit; and if the Nicol's prism is placed with its principal section at right angles to, and immediately before, or immediately behind, the slit, then it is evident that all vertically polarized light will be excluded, and the spectrum visible will be due solely to the unpolarized light of the corona.

Examine the thermal properties of the coronal spectrum. For this purpose a delicate thermoelectric pile and galvanometer will be required; and as glass is comparatively opaque to heat waves, a reflecting telescope should be employed, the lenses of the spectroscope should be of rock salt, and the dispersion should be obtained, either by means of a rock-salt prism, or by means of a reflecting diffraction grating.

As totality approaches, observe with an integrating spectroscope what rays fade away and then brighten; and observe particularly whether all the Fraunhofer lines are reversed at the instant of totality During totality note the intensity of $C$ relatively to $F$, and the intensity of 1474 relatively to $F$, $C$, $D_3$, and $b$. Bear in mind that the relative intensities may change during totality, and pay attention to that point. Watch for the reversal of the Fraunhofer lines at the close of totality. Instrument, $B$ or $E$.

Examine the appearance of the eclipsed sun with a meteor spectroscope, having neither collimator nor slit. So far as the corona is monochromatic it will be distinctly seen notwithstanding the prism, while those portions of it which shine only by reflected sunlight will be indistinct, their light being dispersed. The same object may be attained to some extent by merely looking at the corona through an ordinary prism, or through a direct-vision combination. Instrument, $F$.

*Photography.*—The duration of totality is so brief that it is exceedingly desirable to photograph the spectra then seen, in order that they may afterward be examined at leisure. Attempts to accomplish this have been made with two very different kinds of apparatus, but thus far with only moderate success. One variety of the apparatus, known as the prismatic camera, belongs to the class of slitless spectroscopes, and consists of a photographic camera before whose objective a prism is placed, the whole being equatorially mounted and driven by clock-work. The adjustment of the prism to minimum deviation is effected by the aid of a collimator temporarily placed before it; but on some accounts it would be advantageous to dispense with the prism entirely, and to use in its stead a diffraction grating placed behind the objective. The photographs produced by this instrument exhibit a dispersed series of spectral images of the corona and prominences, each image being due to light of a definite degree of refrangibility. The other variety of apparatus consists of a spectroscope of the ordinary form, except that it is provided with a small camera instead of a reading telescope, the whole being either attached to an equatorially-mounted reflector of short focus, or else so arranged upon a table that an image of the sun may be thrown upon its slit by means of a suitable heliostat and lens. With the prismatic camera, pictures were obtained during the eclipse of April 6, 1875; but thus far it is believed that no results have been got from the other form of apparatus. The reason for this is not apparent, because the brightness of the image depends mainly upon the intensity ratio of the lens by which it is depicted upon the sensitive plate, and there is no reason why this ratio may not be the same in the two forms of apparatus. Theoretically, the slitless form should work slightly the quickest, because it has fewest optical surfaces; but, as an offset to this, there is great difficulty in determining the wave-lengths of the somewhat complicated series of spectral images which it produces.

A careful search in the library of the Naval Observatory failed to reveal any data upon which to found an estimate of the value of the exposure constant for the 1474 line. Under these circumstances, if spectroscopes with slits are used it will probably be best to expose a single plate during the whole three minutes of totality; and as prisms give brighter spectra than diffraction gratings, the former will be preferable for any moderate degree of dispersion; unless, indeed, the object is to photograph the ultra violet rays. If prismatic cameras are employed, the exposure constant should probably be from 0.234 to 0.469—at least, these were the values used by the English expedition to Siam in their work on the eclipse of April 6, 1875.

## SECTION VII.—POLARISCOPIC OBSERVATIONS.

Observations of the polarization of the corona were first undertaken at a time when it was still uncertain whether the corona was a true solar phenomenon, or was in some way produced by the earth's atmosphere; and their primary object was to decide between these two hypotheses. Since then the spectroscope and photography have placed the solar origin of the corona beyond doubt; but the polariscopic observations have proved so delicate that the results hitherto obtained from them cannot be regarded as final. Besides, the details of the polarization, if sufficiently well defined, may reveal something of the condition of the matter emitting the coronal light; and if photography can be substituted for eye-observations, information may even be obtained concerning regions further from the sun's surface than any of which we have at present cognizance.

*Description of Instruments.*—The polariscope has assumed many forms, but those best adapted to the investigation now under consideration may be briefly described, as follows:

*Instrument A.*—A Savart polariscope, consisting of a compound quartz plate and a Nicol's prism or tourmaline; preferably the latter. This instrument indicates the presence of polarized light by exhibiting a series of bands, either parallel or perpendicular to the plane of polarization; and it is so delicate that these bands become visible whenever the polarized light amounts to so much as one per cent. of the total illumination. To discriminate between the plane of polarization and a plane perpendicular to it, we must know whether the central band is black or white, and as that is a difficult question when the bands are faint, it will be desirable to make a small addition to the apparatus. The Savart polariscope being attached to the eye-piece of a telescope in the usual way, a light-colored tourmaline should be inserted in the eye-piece in such a position as to cut off a small segment of the field of view; the axis of the tourmaline being parallel to the edge or chord of the segment, and the bands being perpendicular to that chord. Of course the tourmaline must be so placed as to be distinctly visible through the eye-piece, and it must also maintain its position relatively to the Savart when the eye-piece is rotated. With this arrangement, the plane of polarization will be determined by noting whether the bands continue across the tourmaline or break off at its edge.

*Instrument B.*—A pair of Babinet's quartz compensating wedges, placed at the common focus of the objective and eye-piece of a telescope, the eye-piece being provided with an analyzer consisting of a Nicol's prism or tourmaline. This instrument indicates the presence of polarized light by exhibiting a series of bands, which make an angle of forty-five degrees with the plane of polarization. It nearly equals the Savart in delicacy, and excels it in the facility it affords for determining the plane of polarization, because the position of the middle band can be marked upon the quartz, and then a glance suffices to show whether it is black or white.

*Instrument C.*—A biquartz, arranged in the same manner as the Babinet's wedges just described. This instrument indicates the presence of polarized light by the coloration of the quartz; which, however, does not become perceptible until the polarized light amounts to from 5 to 15 per cent. of the total illumination. If the biquartz is turned till its two halves are colored alike, both being of the purple hue known as the sensitive tint, or tint of passage, then the line of junction of the two pieces will be in the plane of polarization, and as a very slight rotation suffices to destroy the uniformity of the tint, this position can be found with great exactness. Thus it appears that a biquartz is inferior to the instruments *A* or *B* in sensitiveness to feebly polarized light, but superior to them in the accuracy with which it will determine the plane of polarization.

*Instrument D.*—An Arago polariscope, which consists either of a double-image prism and a plate of quartz cut parallel to the axis, or of a double-image prism and a plate of selenite. It will be most convenient to mount the prism and plate of quartz, or selenite, in a brass cell, which can either be used alone, or can be inserted in one end of a tube about a foot long which contains at its opposite extremity a diaphragm of such size that the two images of it, seen on looking through the polariscope, may appear just in contact with each other. This instrument indicates the presence of polarized light by the coloration of the field of view, and has about the same degree of sensitiveness as the biquartz, but it is not well adapted to the exact determination of the plane of polarization.

*Instrument E.*—An achromatic double-image prism.

*Instrument F.*—A Nicol's prism, or a tourmaline. These instruments, as well as *E*, are specially valuable when the light is faint, but they will scarcely indicate the presence of polarization unless its amount is at least 10 per cent. of the total illumination.

*A Polarimeter* will be necessary to determine the percentage of polarized light present. It should consist of four plates of glass, turning upon an axis lying in a plane parallel to their surfaces and at right angles to their length, and carrying an index moving over a graduated circle arranged to show the angle through which the plates are turned. The object whose light is to be tested is viewed through the polarimeter with a Savart's polariscope, so set that the bands seen in it are parallel to the axis of rotation of the plates; then the whole instrument is turned until the bands are perpendicular to the plane of polarization of the light to be tested; and, finally, the plates are rotated until the bands disappear. When this happens, the amount of polarization present is just equal to that produced by the plates, the amount of which is known from the reading of the index.

With the exception of *D* and *E*, which are best used in the hand, all the above-described polariscopes should be attached to telescopes provided with positive eyepieces magnifying about twenty diameters, and giving pencils of light sufficiently large to fill the pupil of the eye completely. Before using the telescopes their lenses should be examined by the aid of polarized light, to make sure that they are free from strain, and after putting the instruments together again special care must be taken to see that all reflections are stopped out in the tubes. Any neglect of these precautions may lead to serious errors.

*Points to be investigated.*—On the whole, the observations hitherto made seem to lead to the conclusion that the corona is strongly polarized in planes passing through the sun's center; or, in other words, radially; and that upon this radial polarization there is superposed a much feebler vertical polarization, which is nearly uniform over the whole corona. The radial polarization is probably due to reflections taking place within a few hundred thousand miles of the sun's surface, and Professor Pickering has shown that if it is produced in accordance with Fresnel's theory it should be strongest in the outer corona, and should diminish to nothing close to the sun's limb. The vertical polarization is probably due to the action of our own atmosphere.

Theoretically there should be no difference between the indications of an Arago polariscope with a quartz plate and one with a selenite plate; and upon ordinary objects the results given by these two forms of the instrument are certainly identical; but Professor Pickering thinks that his observations upon the eclipses of August, 1869, and December, 1870, show that there is some peculiarity about the corona which causes its light to behave differently toward a quartz plate from what it does toward a selenite one.

From these considerations, it will be gathered that the questions demanding special attention are the following:

1. Is the light of the corona polarized radially or vertically, or in both these ways?

2. If radial polarization exists, does it extend with unabated intensity down to the sun's limb, or does it continually diminish, and finally vanish, in passing from the outer corona inward to the chromosphere?

3. What is the percentage of polarized light present?

4. In examining the corona with an Arago polariscope, do the results given by an instrument with a quartz plate differ from those given by an instrument with a selenite plate?

*Methods of Observation.*—If a beam of light polarized in a vertical plane is examined through any of the polariscopes described above, the appearances presenting themselves as the instrument is rotated will be as indicated in the following table:

| Azimuth of Principal Section of Polariscope. | Condition of Bands seen in Savart Polariscope. | Condition of Bands seen in Babinet Wedges. | Colors seen in a Bi-quartz. | Colors of the Two Images in an Arago Polariscope. | Brightness of the Two Images seen through a Double-Image Prism. | Brightness of Light seen through a Nicol or Tourmaline. |
|---|---|---|---|---|---|---|
| ° 0 | Maximum. | Disappear. | Violet. | Red. Green. | Maximum. Minimum. | Maximum. |
| 45 | Disappear. | Maximum. | Green, Red. | Colorless. | Equal. | |
| 90 | Maximum. | Disappear. | Yellow. | Red, Green. | Max., Min. | Minimum. |
| 135 | Disappear. | Maximum. | Green, Red. | Colorless. | Equal. | |
| 180 | Maximum. | Disappear. | Violet. | Green. Red. | Minimum. Maximum. | Maximum. |
| 225 | Disappear. | Maximum. | Red, Green. | Colorless. | Equal. | |
| 270 | Maximum. | Disappear. | Yellow. | Green, Red. | Min., Max. | Minimum. |
| 315 | Disappear. | Maximum. | Red, Green. | Colorless. | Equal. | |

Of course the changes indicated in the table take place, not suddenly, but gradually and continuously, as the polariscope is rotated. With the Savart and Babinet instruments, if the bands are black centered at the first and third maxima they will be white centered at the second and fourth maxima, or vice versa. In describing the action of the biquartz and Arago instruments the colors red and green are mentioned; but it must be clearly understood that the two complementary tints actually given by any instrument are determined solely by the thickness of the quartz or selenite plate employed.

At first sight the table seems to indicate that no two polariscopes are governed by the same law; but such a conclusion would be a mistake. They all exhibit two perfectly distinct test phenomena, or reactions; as for example, bands with a white center and bands with a black center, or a given tint and its complementary, or a maximum and a minimum of light; and if in any given position of the principal section of the polariscope one of these reactions is at a maximum, then the same reaction will also be at a maximum when the position of the polariscope is 180° different, while for positions 90° or 270° different the other reaction will be at a maximum.

Now, supposing the eclipsed sun to be viewed through a telescope armed with a polariscope, let us apply the principles just enunciated to the solution of the first of the questions proposed above. The corona will be brought to the center of the field of view, and the polariscope will be turned until it exhibits a maximum of one of the reactions. Then, if the polarization is vertical, this reaction will be exhibited over the whole corona; but if the polarization is radical, the corona will appear divided into four quadrants, the first and third of which will exhibit one of the reactions while the second and fourth will exhibit the other. Next, the polariscope will be slowly rotated, while the corona is kept steadily in the center of the field of view. Then, if the polarization is vertical, the first reaction will gradually fade out and be replaced by the second, which in its turn will be again replaced by the first, and so on, as indicated in the table above; but if the polarization is radial, the two reactions exhibited in the alternate quadrants of the corona will not undergo any change, except that they will rotate with the polariscope.

The observations necessary for the solution of the second question proposed above may be made with almost any polariscope by noting whether or not the reactions just described continue undiminished quite down to the moon. It is probable, however, that the results obtained from a biquartz will be most satisfactory. It should be set so that the normal to the moon's limb makes an angle with the line of junction of its two halves sufficiently large to cause their colors to contrast vividly. Then the observation will consist in noting whether this contrast continues with unabated force all the way to the moon's limb, or whether it gradually diminishes, and finally fades out.

The amount of polarized light present in the corona can only be determined by measurements with the polarimeter, the method of using which has been already described.

In dealing with the fourth question proposed above, it is desirable that the two Arago polariscopes employed should give as nearly as possible the same colors. The method of proce durewill be as follows: Firstly, the corona will be examined through one of the polariscopes, furnished with its tube and diaphragm, and the reactions, both upon the corona and upon the sky, will be noted; secondly, the tube and diaphragm will be removed from the brass cell containing the prism and plate, and the corona will be again observed. In this case, the images of two portions of the sky, distant from each other about two and a half degrees, will be superposed, thus eliminating their polarization, and therefore it will only be necessary to note the reaction upon the corona itself; thirdly, all the observations will be repeated with the other polariscope.

In order to familiarize themselves with the appearances likely to be presented during the eclipse, intending observers will do well to practice beforehand upon an artificial corona radially polarized, such as has been described by Professor Pickering in the United States Coast Survey Report for 1870, page 167. The polariscopes which seem likely to give the best results are Babinet's compensating wedges and the biquartz.*

*Photography.*—Among the various forms of polariscopes the Nicol's and double-image prism seem almost the only ones available under the peculiar conditions which surround attempts to photograph the corona. If a Nicol is employed, its aperture will probably be comparatively small, and on that account it should be placed behind the photographic objective, and near its focal point; but if a double-image prism is used, its aperture should be sufficient to permit its being placed immediately before the objective. On the whole, the simplest form of photographic polariscope, and that most likely to yield satisfactory results, seems to be an ordinary camera, provided with a photographic objective having an intensity ratio of about one-eighth and a focal distance of about twelve inches, immediately before which is mounted a double-image prism with a clear aperture equal to that of the objective. With this apparatus wet collodion plates of ordinary sensitiveness should be exposed about thirty seconds, and after each exposure the prism should be rotated thirty degrees, the motion being in the same direction as that of the hands of a watch.

## SECTION VIII.—PHOTOMETRIC OBSERVATIONS.

As there is now reason for suspecting great variability both in the size and in the brightness of the corona, exact photometric observations during total solar eclipses become of much importance; and in making them, attention may be directed either to the general illumination of the atmosphere, or to the quantity of light emitted by the corona.

For investigating the first of these points the apparatus devised by Prof. J. R. Eastman, and used by him in observing the eclipse of August 7, 1869, will be found

* For further details consult the following: Report on Observations of the Total Solar Eclipse of December 22, 1870. By Prof. E. C. Pickering. U. S. Coast Survey Report for 1870, p. 165. List of Observations on the Polarization of the Corona. By Prof. E. C. Pickering. Journal of the Franklin Institute, 1871, vol. 61, p. 58. Instructions for Observers at the English Government Eclipse Expedition, 1871. Nature, 1871, vol. 5, p. 18. Suggestions to Observers of the Solar Eclipse of December next. By A. C. Ranyard. Nature, 1871, vol. 4, p. 327. English Government Eclipse Expedition, 1875. Instructions to Observers. Nature, 1875, vol. 11, p. 352.

very satisfactory.* On that occasion the light during totality proved about equal to that on a clear moonless evening, at the time when the sun has sunk so far below the horizon that third magnitude stars are just becoming easily visible.

For measuring directly the amount of light emitted by the corona, some form of Bunsen's photometer will be required, and probably that adopted by Professor Pickering will be found as convenient as any.† It consists of a box 9 inches wide, 18 inches high, and 6 feet long, within which a lighted candle can be moved backward and forward by means of a long projecting rod. One end of the box is covered by a piece of thin white unruled writing paper, in the center of which is a greased spot about half an inch in diameter. The observation consists in pointing the box toward the corona, so that the rays of the latter may fall squarely upon one side of the paper while those of the candle fall upon the other; moving the candle by means of the rod until the greased spot disappears; and then marking upon the rod by a stroke of a lead-pencil the distance between the candle and paper. A number of disappearances should be observed, and after totality is over the distances between the marks and the candle must be measured and recorded. During the eclipse of December 22, 1870, Professor Pickering found that, on the average, the candle had to be 18.5 inches from the paper to make the spot disappear, but the observations were much embarrassed by flying clouds.

Owing to the difference of color between daylight and candle-light, no adjustment of the candle will suffice to make the greased spot disappear completely; it can only be reduced to a minimum of visibility.‡ Care must be taken to have the interior of the photometer-box painted dead black with a mixture of lamp-black, shell-lac, and alcohol. The candle must be of wax or sperm, of the kind known as six to the pound, and experiments must be made to determine exactly how many grains it burns per hour; or, better, the candle may be sent to the Naval Observatory and these experiments will be made there.

Persons who are inexperienced in photometry should have some practice previous to the eclipse. The necessary apparatus will be, two lighted candles placed three or four feet apart, and a sheet of thin white writing-paper, having in its center a spot half an inch in diameter, rendered transparent by the application of a little grease. The practice will consist in holding the paper in, and at right angles to, the line joining the two candles, and then moving it backward or forward until the greased spot disappears, which it will do when the illumination on the two sides of the paper is exactly equal. When this experiment has become quite familiar, the photometer which is to be employed during the eclipse should be taken into a room dimly lighted by daylight, and the intensity of the illumination should be repeatedly measured by making the spot disappear as completely as possible, and then noting the interval between the candle and the paper. Excellent practice may also be had by measuring the intensity of twilight in the early evening.

---

* Reports on Observations of the Total Eclipse of the Sun, August 7, 1869, Appendix II to the Washington Observations for 1867, p. 100.

† See United States Coast Survey Report for 1870, p. 172.

‡ The greased spot can be made to disappear completely by using paper which is blue on one side and white on the other; but this cannot be recommended, because it introduces serious theorectical difficulties.

## SECTION IX.—THERMO-ELECTRIC OBSERVATIONS.

The principal points to be investigated are the total amount of heat emitted by the corona and chromosphere, and the relative temperatures of the prominences and of different parts of the corona. The galvanometer employed should be very sensitive, and it will be advisable to have some small resistances which can be inserted in the circuit to reduce the deflections of the needle in case the currents given by the pile prove unmanageably strong.

For investigating the first of the points mentioned above, an ordinary thermo-pile, provided with the usual conical reflector, will be the best instrument; and it should be used by pointing it to the corona without the intervention of any condensing lens or mirror whatever. After measuring the heat in that way, its so-called quality may be tested by noting the effects produced by the interposition of screens of glass or other transparent material of various known thicknesses.

For investigating the relative temperature of the prominences, and of different parts of the corona, a thermopile placed in the focus of a telescope, which should be a reflector if possible, will be required; and as we are ignorant of the degree of heat to be expected, the pile should not be too delicate. A very suitable one can be made by taking a piece of iron wire about six inches long, soldering to each of its extremities pieces of german-silver wire, and then bending the whole into the form of an M, the solderings being at the points forming the top of the M. The wire should not be more than one or two hundredths of an inch in diameter, and the whole should be so mounted in a piece of cork that the junctions may project only a short distance above its surface. The currents given by this pile will be due solely to the differences of temperature of its two junctions, and, consequently, the galvanometer will be insensible to any flow of heat which affects both junctions alike. The mode of using the instrument will be to place one of its junctions upon the moon, or upon a prominence, and the other upon the point whose relative temperature is to be measured. One of the most important points to be ascertained is how much more heat comes from the prominences than from the neighboring portions of the corona.

As soon as possible after the totality is ended, the constants of the apparatus should be most carefully determined, so as to furnish the means of converting the galvanometer readings into ordinary thermometric degrees.

o

Plate 1.

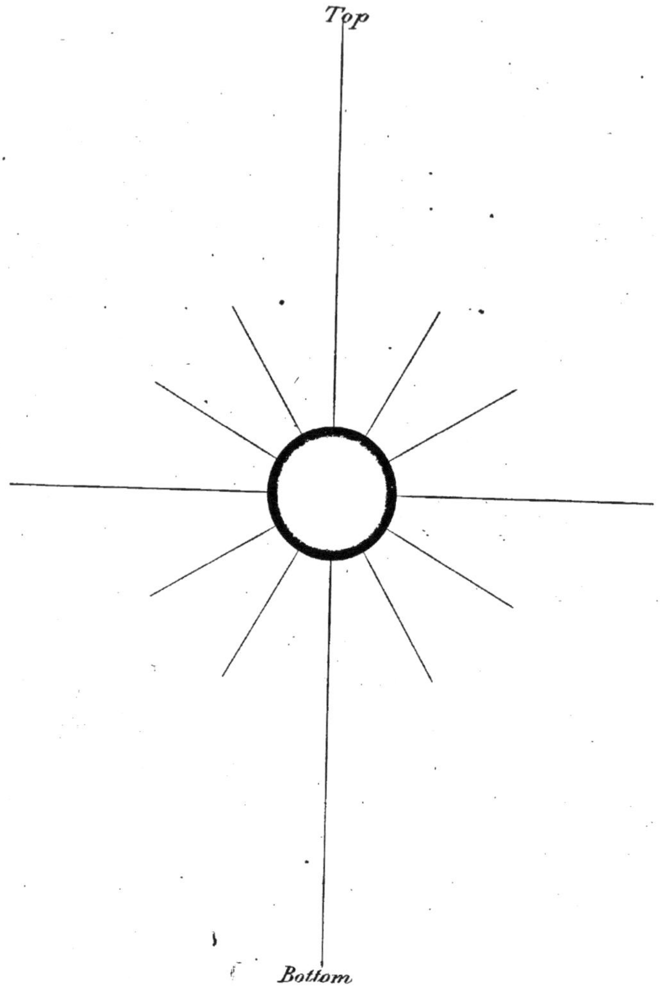

Plate II,

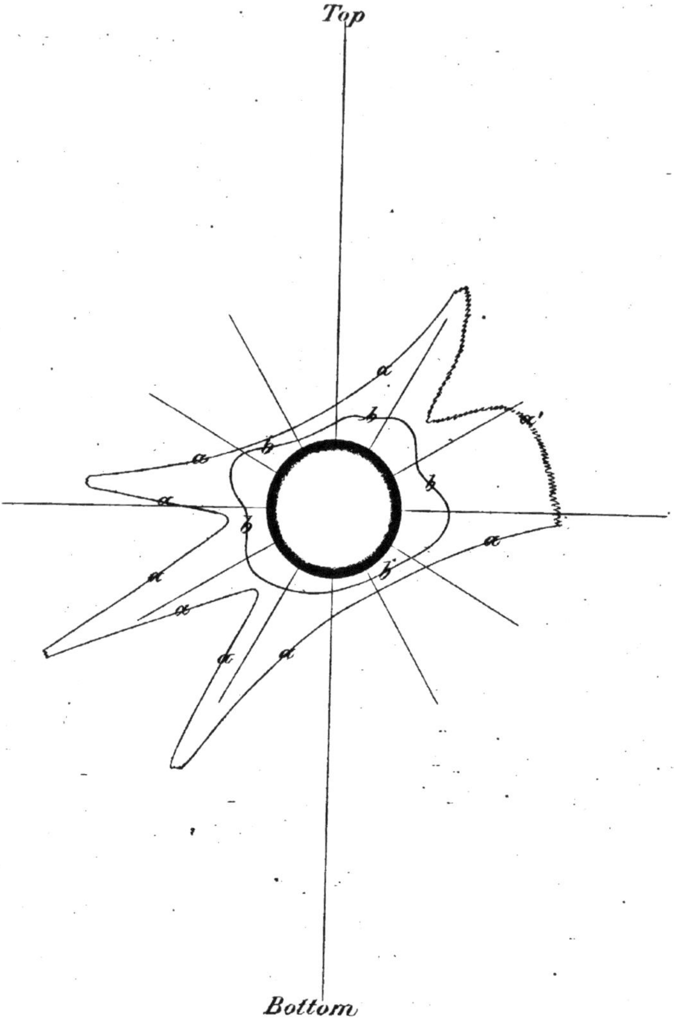

Spotted something that isn't quite right?

Let us know and we'll fix it for future reprints. And as a thank you for your diligence, we'll send you a free Sothis book of your choice.

publisher@sothis.press

This page intentionally left blank

Except of course for that line above. And this one.
And whilst we have your attention, perhaps you'd
like to follow us on Instagram @sothis.press. Go on.

Made in the USA
Monee, IL
08 November 2026

68103657R00025